Could Current Science Explain Paranormal Phenomena?

Introduction to Concepts of "Broadcast-Reception" and "Broadcast-Storage-Retrieval" of Information and Energy *via* the "Human Wireless Senses"

By Malika Ammam, PhD

One way or another, you likely heard the terms "paranormal", "supernatural" or "sixth sense", whether through the longest-running science fiction series "The X-Files" or the 1999 American supernatural horror film "The Six Sense". These interchangeable terms, oftentimes referred to as paranormal, encompass phenomena that could not be explained by the current scientific view of laws of nature. Paranormal phenomena include dreams, telepathy, telekinesis, reincarnation, divination, prophecy, haunting, and curses, among others. Paranormal topics are often ignored by the scientific community due to lack of solid evidence, and any attempt to provide rational explanations usually results in raised brows and smirks. Of course, most reported phenomena are fraud, self-delusion and/or self-deception but a faction of events might hold some truth and merit investigation. The truth of the matter is that a little is so far provided to rationalize paranormal phenomena due to their extreme complexity. Besides, existing explanations are often phrased in vague terms, such as "psychic forces", "human energy fields", and even propositions speculating the existence of a fifth physical force in addition to the four known forces being gravity, electromagnetism, weak and strong forces. But do we really need these ambiguous terms to comprehend at least part of the paranormal events or could current science provide some answers?

We all dream while asleep. They are often a source of inspiration, creativity, and coming up with new ideas. We might say a dream is a series of images and sensations that makes a story during sleep. The events occurring in dreams might seem just like those happening in an ordinary day or they can be weird, surreal or even frightening. Dreams mostly occur during the rapid eye movement (REM) phase of sleep, when the brain activity is similar to that of a waking state. During dreams, we generally have no control over what we see, hear or feel, and we quickly forget the content soon after we wake up. However, we can have the power over what happens in dreams through lucid dreaming, when we become aware of the fact that we are actually dreaming.

Throughout history, dreams have been studied by philosophers, religious scholars, and scientists. But despite this, whether or not dreams have a purpose remains shrouded in mystery. Ancient peoples considered dreams as religious or supernatural messages provided by divine spirit reals or gods. For instance, ancient Sumerians believed that dreams possess prophetic power and could predict the future.[1] As a result, dreams played an important role in shaping the Sumerian civilization since they used dream interpretation to grasp and understand the divine prophecy. For example, king Gudea, whose reign spanned from 2144-2124 BC, rebuilt Ningirsu temple as told in a dream. For ancient Egyptians, dreams were seen as a blessing and divine revelation,[2] where priests often visited sanctuaries to sleep on "dream beds" to help them receive more vivid dreams containing words of advice or comfort from the gods. Similarly, ancient Greeks slept in temples or at shrines believed to enhance their abilities to receive prophecies (or even warnings) from the god of dreams, Morpheus.

Others thought of dreams differently. Ancient Chinese believed that the soul leaves the body during sleep, and dreams are experiences felt while roaming around the earth, which then projected back to the human body as dreams.[3] The Hindus Indians consider the soul in three states: wake, sleep, and dream state.[4] Buddhists believe that if multiple peoples have the same dream, this would indicate a prophecy, and may even transcend time since the future Buddha may have the same dreams as previous ones. Indigenous American and Mexican civilizations consider dreams as a way of visiting and contacting their ancestors.[5]

The Greek philosopher Hippocrates (469-399 BC) thought that dreams originate from the soul projecting images received during the day. Carl Jung considered dreams as messages to the dreamer bringing revelations that might uncover and help to resolve emotional or religious issues and dreads.[7] Frederic W. H. Myers, one of the founders of the Society for Psychical Research,[8] proposed that two or more humans could transfer information in the form of words, images or sounds from one mind to another without ever making physical contact or talking to each other. He called this process "dream telepathy", and believed that information transfer between minds could even be possible to achieve over wide distances. The dream telepathy was first psychoanalyzed by Sigmund Freud in his paper "Dreams and Telepathy", which was later reproduced in 1953 as a book entitled "Psychoanalysis and the Occult". In the 1940s, Jule Eisenbud, Geraldine Pederson-Krag and Nandor Fodor reported cases of dream telepathy. Stanley Krippner and Montague Ullman from Maimonides Medical Center (Brooklyn, New York) conducted research on dream telepathy and concluded that telepathy is real.[9]

While dream telepathy takes place during sleep, information might likewise be transferred telepathically while awake. In October 1937, the author Harold Sherman and explorer Hubert Wilkins did some experiments on telepathy for a period of five and a half months. The experiment consisted of visualizing a mental image of daily events or thoughts experienced by both researchers, one located in New York and the other in the Arctic. When Sherman and Wilkins compared their diaries, they found over 60 percent similarity.[10] Today, parapsychology describes various forms of telepathy.[11] The first is latent telepathy characterized by the delay between the information transmitted and received. In retrocognitive, precognitive and intuitive telepathy, a person would be able to communicate information about the past, present or future to another person. This is the most varied form of telepathy and common cases would be touching an object and gleaning information about its past, reading someone's mind, or seeing future events before they happen. Emotive telepathy is when a person remotely alters someone else emotional state through the transfer of sensations. In superconscious telepathy, one could tap into the collective knowledge of human civilization.

In addition to the transfer of information through dreams and telepathy, energy can also be transmitted remotely over long distances *via* psychokinesis or telekinesis. The word "telekinesis" was first coined by the Russian psychical researcher Alexander N. Aksakof in 1890,[12] and "psychokinesis" was first used by the American author Henry Holt in his book of 1914 "Cosmic Relations".[13] Using telekinesis, a person could basically influence a physical system without actually being in contact with it. The truth of the matter is that most reported telekinesis phenomena are built on trickery, special effects or poor experimental design. However, a small portion might hold water. For instance, numerous Indian's "godmen" exhibited telekinesis abilities in public, in presence of controls to prevent trickery.[14] According to the Menninger Foundation, the yogi skilled Swami Rama was skilled in controlling his heart functions and allegedly succeeded to telekinetically move knitting needle twice from a distance of five feet.[15] The Russian psychic Nina Kulagina, mentioned in the U. S. Defense Intelligence Agency report of 1978, was filmed performing telekinesis shows in several black-and-white short films.[16]

But how do such phenomena take place and what would be the force at play? Within the field of parapsychology, this transfer of information or energy through the mind at long distances is considered as a form of extrasensory perception (ESP), which literally means outside the known senses of sight, hearing, touch, smell, and taste. Some suggest that these paranormal experiences occur solely through the mind. Others argue that they ascend from the heart, gut or something mysterious. Whatever this extrasensory perception could be, some believe that it is independent of the limits of space and time. But is this really true?

Our existence consists of reactions to our surroundings. We experience the world around us through our senses or "sensors", of which by reacting to stimuli provide data for perception. The main five traditionally recognized senses are sight (vision), hearing (audition), taste (gustation), smell (olfaction), and touch (somatosensation). To prepare for a proper response, our brains are constantly collecting and analyzing information induced from our sensory organs. These sensors are made of cell receptors sensitive to changes in light, sound, smell, and touch, among others. The gathered information is then converted into electric signals, called "impulses", which are forwarded to the brain through the nerve cells or "neurons" for interpretation. Impulses produced from different parts of the body are directed to various regions of the brain for analysis. For instance, smell-induced impulses are processed by the gustatory cortex of the brain while sound-generated impulses are examined by another part of the brain known as the auditory cortex.

Sight or vision is realized through our eyes, which are capable of detecting images every time visible light bounces into the eyeball. The outer layer of the eyeball called "retina", comprises rods and cones photoreceptors, which engender electric nerve impulses upon reception of light. Rods are sensitive to light but cannot differentiate color while cones can differentiate color and operate under dark. Vision received from the left visual field is transmitted from the right side of each retina through the optic nerve before reaching the visual cortex of our brain for reading.[17] People who are unable to see are blind. This might be caused by the destruction of the retina, malfunction of the optical nerve connecting the eye to the brain, or diseases like stroke.

Hearing or audition is accomplished through our ears, made up of three parts: the outer ear, a middle ear which contains eardrums or the membrane that vibrates upon reception of sound, and inner ear. Sound is recognized as vibration, and the inner ear owns mechanoreceptors that turn air motion or vibration into electrical impulses. The induced information then travels through the auditory nerves before reaching the auditory cortex of our brains for analysis. The incapability to hear is known as deafness. This could be induced by aging or damage to the middle ear, inner ear, or the nerves that transmit impulses to the brain.

Taste or gustation of food and other substances is distinguished through the tongue. The human upper surface of the tongue is composed of cell receptors called "taste buds or gustatory calyculi". These senses are able to detect basic tastes like sweetness, bitterness, soreness, saltiness, and yumminess. The assembled information then moves through appropriate nerves before getting to the gustatory cortex of the brain for assessment. Injury to the bud cells, gustatory nerves or gustatory cortex of the brain results in the failure to taste, known as ageusia.

Smell or olfaction takes place through the noise. As we breathe, different particles present in the air bond to our smell receptors found in the nose ceilings or nasal cavity. The induced-impulses are then conducted to the olfactory complex of the brain by nerve cells. Our noses work closely with our taste receptors to let us taste different flavors as opposed to the five main tastes detected by our tongues. Unlike other neurons which can not regenerate, olfactory receptor neurons do this on a regular basis. The incapability to smell is known as anosmia and could result from damage to the olfactory receptors, the nerve cells, or olfactory complex of the brain.

Touch or somatosensation is an unusual sense because it is the only conventional sense not found around the head region. The sense of touch arises in most regions of the skin covering our bodies. The touch receptors found on the skin, including hair follicles, are sensitive to pressure. Impulses induced by touch and vibration from the skin and joint position are then channeled by nerve cells to the brain for evaluation. There are other subtler senses that work with our sense of touch. Peoples with tactile anesthesia are unable of feeling anything when touched.

But are these the only senses that humans actually possess or there is more? For decades, we are taught that humans have five senses but the truth is that we can sense a lot more things beyond the conventional senses. This may be simple as hunger or complex as the abstract of time. These unconventional senses are there and taken for granted but play important roles in survival. For every stimulus, our bodies have receptors capable of reacting to it and take decisions as to instructions to send to the rest of the body.

Pain and temperature count as sensations. The aptitude to feel physical pain is known as nociception, often caused by damage to the nerve or tissue. The receptors of pain are localized in areas of cutaneous (skin), visceral (body organs), and somatic (joints and bones). Pain plays an important role in warning us of bigger threats, such as hot water that can burn our skin or body stretching that may dislocate our joints. Pain is traumatic enough to remind us to stay vigilant and not repeat the same mistakes. Our faculty to sense heat works very closely with our sense of touch and accomplished by cells named "thermoreceptors". Actually, there are separate receptors responsible for feeling hot and cold temperatures. The receptors detecting the internal body temperature are recognized as "homeostatic" thermoreceptors, which are different from those we have on the skin for sensing pain, cold or heat.

For movement, humans own two kinds of senses: balance or equilibrioception and kinesthetic or proprioception. Equilibrioception helps us to maintain body posture by sensing the movement, direction, and acceleration. This is undertaken by the vestibular labyrinth system found in the inner ear meanwhile involved in hearing. Proprioception, on the other hand, enables us to sense relative positions of different body parts. Impaired kinesthetic and balance senses would cause troubled bodily movements.[18]

A number of senses closely related to internal organs also exist. An internal sense is any sensor that is able of reacting to changes in our bodily functions.[19] Usual sensations like hunger are controlled by the brain section called "hypothalamus".[20] Likewise, there are chemoreceptors monitoring chemical changes in our bodies, such as excess carbon dioxide that makes us feel suffocated, lack of water causing the feeling of thirst, and hormone-related activities like high blood sugar that triggers diabetes. Other organ-related sensors are present too, including those that detect vomiting and gag reflex in the digestive and respiratory tracts, as well as the fullness of urinary bladder and rectum. Equally, humans have sexual stimulation senses that could stem from the touch of the erogenous zones or simply triggered by sexual hormones.

To sustain awareness, humans possess senses of time or chronoception and recognition memory. Chronoception is mainly responsible for tracking time and maintaining the notion of time intervals.[21] Depending on the duration, the sense of time could be classified into short term patterns (such as heart rate, sleep, and hunger) or long term patterns (such as menstrual cycle).[22] In addition, our central nervous systems are adept of sensing the notion of subjective time, perceived as relative feeling and action-linked duration that makes it feel slower or faster than it actually does. People with schizophrenia have lost this ability, making them feel like machines. On the other hand, recognition memory senses are responsible for the feeling of familiarity and recollection.[23] We can experience a sense of familiarity without any recollection, such as in the case of "Deja vu". Studies showed that mice with the wounded perirhinal cortex, the part of the brain responsible for sensing familiarity, are powerless of differentiating novel images from familiar ones.

Compared to humans, animals have much stronger and advanced senses, allowing them to adapt to extreme surroundings, such as low-light habitats or simply to sustain the life of their species. Some animals modify their primary senses to assist them to navigate by estimating positions. For instance, sharks follow the first nostril that detects a smell to track its source.[24] They combine this information with the appearance timing of the smell to determine its direction. Bats and cetaceans like dolphins and whales custom their hearing organs for echolocation to determine the position of surrounding objects by analyzing the sound waves they reflect. Many invertebrates, especially aquatic mollusks, and medusas own a special sense called "statocyst", which helps them to track their movement when they speed up and provide balance information.

Certain animals enhanced their sense of vision to help them navigate or protect themselves. Cats can expand their pupils to see in low light environments. They also possess a membrane in their eyes "tapetum lucidum" that might reflect admitted light and enhance image resolution. Some animals like birds, bees and dragonflies are tetrachromats, meaning that they could see in the ultraviolet light. Mantis shrimps have twelve different kinds of color receptors while humans and other mammals possess only three and two, respectively. Also, they are competent in converting circularly polarized light into linear light to enhance their vision and help them find a suitable mate. Cephalopods, such as squid, octopus, and nautilus have only one color sensor in their eyes, so they are technically colorblind. The molecule "opsin" on their skin also helps to detect light of different wavelengths and determines what color to pick up in each environment. This faculty comes handy when they do camouflage, matching their own color to their surroundings.[25]

Other animals added senses to sustain life, either for hunting or mating. Snakes and vampire bats are skilled in sensing the body heat of their prey. They possess sensors to detect infrared light (in the face for snakes and nose for vampire bats).[26] Flies and butterflies developed senses in their feet to taste flavors of anything they land upon. Catfish have taste senses all over their bodies, enabling them to taste anything that surrounds them, including the chemicals present in the swimming water.[27] Insects also have olfactory or smell detectors in their antennas. Others animals like reptiles and mammals own a special smelling organ located in the mouth cavity. This organ is handy for detecting non-volatile smell molecules like pheromones for marking territories and trails or for mating.[28]

More fascinating, some animal species have senses that detect electric and magnetic fields, known as "electroception" and "magnetoreception", respectively. Numerous fish species and sharks are adept of sensing changes in electric fields that surround them, and others could generate weak electric fields over their bodies.[29] These abilities to acquire and create electric fields are convenient for social communication. Dolphins are gifted in detecting electric fields thanks to their electroreceptors located in the face areas, such as nose, mouth, and jaw.[30]

They may well sense electric field intensities as low as those induced by contracting muscles of prey to locate them even in murky seawater. Insects like spiders can sense electric fields to guide them in setting a suitable time for extending their webs.[31] Most birds custom the Earth's magnetic field for directional awareness and navigation during migration.[32] Cattle employ magnetoception for north-south direction alignment.[33] Magnetotactic bacteria might easily produce miniature magnets inside themselves and orient relative to the Earth's magnetic field.[34] Electrogenic animals like electric eel may produce power in the form of electric voltage for either hunting or defense.[35] All animals use voltage pulses to transmit information between neurons and muscles.[36] Application of a voltage to the human body induces an electric current through the tissue. Higher voltages engender greater electric currents,[37] with the threshold for perception depending on the supplied frequency and path of the current.[38] Sufficiently elevated currents may lead to muscle contraction, heart fibrillation, tissue burn, and even death.[37]

Electroception, magnetoreception, electrogenic, voltage and current are all phenomena related to electromagnetism, one of the four fundamental forces of nature, the others being strong interaction, weak interaction, and gravity. In the past, electricity and magnetism were considered as two completely separate phenomena. But the fact is that they are two sides of the same coin. Studies carried out by Michael Faraday and Andrè-Marie Ampère revealed that a magnetic field varying with time generates an electric field, and an electric field changing with time induces a magnetic field.[39] More compelling, this phenomenon possesses the properties of an electromagnetic wave. In 1864, James Clerk Maxwell analyzed the properties of the electromagnetic waves and established that such waves would necessarily travel at the speed of light, and concluded that light itself is an electromagnetic radiation.

On the other hand, the electromagnetic radiation is characterized by a frequency of oscillation that determines its energy and wavelength, and radiations of various frequencies form the "electromagnetic spectrum".[40] In the decreasing order of energy, we find gamma rays, X-rays, ultraviolet, visible light, infrared, microwaves, and radio waves.[41] The gamma rays and X-rays are the most energetic radiations, which can be sometimes beneficial for medical, technological, and industrial purposes. Still, uncontrolled exposure to these powerful radiations can be injurious or even fatal. One single photon of gamma radiation has an energy equivalent to 100,000 photons of visible light. Repeated exposure to gamma radiation or X-rays is recognized to cause cancer since the radiation penetrates through the human body and induces mutations in the DNA. The ultraviolet is localized between the lower limit of X-rays and violet color of the visible light. It is responsible for summer sunburns, and high exposure to ultraviolet might harm the skin since it carries sufficient energy to alter the DNA structure but less than X-rays. The visible radiation is the portion of the electromagnetic spectrum where the human eye could see in color. Infrared radiation bears energy below that of the visible red light and mainly used in night vision of surveillance cameras and remote temperature sensing.

Radio waves and microwaves hold the lowest energies of the electromagnetic spectrum and have particular importance in telecommunication and power applications. Nowadays, the use of radio waves and microwaves is so widespread that you probably heard of the numerous technologies which make use of these radiations, such as televisions and radios, mobile phones, computers and laptops, surveillance camera systems, and even wireless power charging gadgets. In the past, nearly all forms of telecommunication and energy transfer were wired.[42] For this reason, underground communications cables and overhead lines were arranged to connect between telephone devices. To talk to another person, a wire connecting both phones is required or there would be no way to transmit and receive calls. The same applies for power transmission. To charge your electric toothbrush or iPhone, your device must be connected to the mural electric outlet. Nowadays, most common forms of telecommunication are wireless, and power transfer from one device to another without the use of the physical link is becoming increasingly frequent.[43] But how does this actually works? In wireless communication and energy transfer, electromagnetic radiation (radio waves and/or microwaves) is transmitted through non-material media, such as the air or vacuum instead of cables, wires or other electrical conductors.[44] This makes it easier and practical to diffuse communication and energy remotely over long distances.

Technically, both wireless communication and power make use of the same electronic components. They need a transmitter, a transmission medium, and a receiver. The transmitter is an electronic device that converts the information or power to be sent into a signal for broadcast.

The produced signal then travels through the transmission medium, which might be air, vacuum or water. Once the signal reaches the destination, it is again reconverted by the receiver gadget into a usable form of information or power. For instance, when you watch a movie on Netflix, listen to music on your favorite radio station or follow road directions provided by your Global Positioning System (GPS), you are actually "streaming" information sent, traveled then converted into a video, audio or video/voice format.[45]

A crucial component to combine with the transmitter and receiver devices is the antenna, which can be defined as a bidirectional artificial sensor since it works in both directions to convert electric signals induced at the transmitter into electromagnetic waves, or transform electromagnetic waves back into electric signals to be analyzed by the receiver. In other words, antennas are made of an array of conductors electrically connected to the transmitter and receiver devices designed to broadcast and receive electromagnetic waves wirelessly over long distances.[46] The first antennas were built in 1888 by the German physicist Heinrich Hertz when he was conducting experiments to prove the existence of electromagnetic waves. Since then, antennas become mandatory components in numerous technologies associated with communication and power transmission, such as walkie talkies, home radios and televisions, mobile phones, laptops, internet Wi-Fi routers, surveillance cameras and so on. But how do antennas actually function? Basically, the data to transmit wirelessly have first to be converted into electric signals by the transmitter device, which will then be passed to the antenna to turn them into electromagnetic waves. Each electric signal passed through the antenna produces a unique electromagnetic wave that travels through air at high speed. Once arrived at the destination, the receiving antenna turns the electromagnetic waves into electric signals and then handed to the receiver that will reconvert them into the original sent information or power.

For instance, a radio communication system, such as a mobile phone, necessitates antennas to transmit and receive data. These antennas, used to emit and pick up radio waves, are made of conducting materials, such as metal rod resonators. The transmitter produces a radio signal by generating an alternating electric current through the antenna. This causes electrons in the metal resonator to oscillate or vibrate, creating an electromagnetic field radiating outwards from the antenna as radio waves. These signals carry unique information (like sound data in a mobile phone), produced by modeling the electromagnetic waves at the transmitting device. After the initial diffusion of the signals and their journey through air, the data-bearing signals are then demodulated at the receiver. In this course, the delivered radio waves hit the antenna, causing electrons in the metal to oscillate in much the same way as the electrons at the transmitting antenna. This generates an alternating current, which is converted by the receiver back to the original information (sound data in a mobile phone). One way to analogize this process is to think of the information as a letter. The modulation process consists of putting the letter in an envelope, the signal transmission through a medium as posting the envelope, and demodulation as opening and reading the letter.

Antennas come in all shapes and sizes. They can be large and bulky or small and concealed. The modest and efficient way to classify antennas is through the radiation pattern of the produced or received electromagnetic waves. Accordingly, we can define two categories of antennas. Omnidirectional antennas emit and accept electromagnetic waves that travel and come from all directions.[47] A common model of this type of antennas is a Wi-Fi router, which radiates radio waves in all directions within the covered area. By comparison, directional antennas induce and collect electromagnetic waves that travel and arise from specific directions. These features boost their performances and lessen interference from unwanted sources. Directional antennas own these properties thanks to additional design features, such as parabolic reflectors or dishes. A typical example of this kind of antennas is the dish TV facing a particular direction to harvest a maximum of the radio waves sent by a television station.

By comparison to artificial antennas, a number of animal and insect species own well developed biological antennas. Arthropods, including insects, crustaceans, and arachnids visibly display natural antennas, often referred to as "feelers". They are found on the arthropod's head and widely vary in the form between species. These antennas hold functions in sensing but how the sensing takes place differs among groups. Some insects like ants also possess antennas. Studies showed that biological antennas could sense various stimuli, such as changes in air motion, touch, heat, vibration (sound), taste, and smell.[48] Some antennas consist of olfactory receptors that bind to odor molecules like pheromones. The nerves associated with these receptors convert information to voltage pulses then forward it to appropriate sections of the brain for analysis and identification of the odor. Monarch butterflies custom their antennas as clocks to determine when to begin sun compass orientation during migration.[49] The African cotton leafworm males employ their antennas for mating, as well as to answer the female mating call.[50] The diamondback moth and giant swallowtail butterfly females make use of their antennas to identify the desirable hosting plants where to deposit eggs. They achieve this by sensing the taste and odor of the plants.[51]

Our progress in artificial antennas and wireless transmission is the result of enthusiastic experimentation with electronics and electromagnetism, trying to replicate senses found in humans, animals, and insects. However, the truth of the fact not all the human biological senses are fully identified. In addition, the mechanisms governing the human senses are so complex that a little still answers the large number of the raised questions. Therefore, one should wonder if humans might actually own biological antennas, operating as "human wireless senses" to explain paranormal experiences, such as dreams, telepathy, and telekinesis, among others. In other words, might the so-called "six sense" or "ESP" be simply "human wireless senses", able of broadcasting and receiving information and energy, just like artificial antennas applied for wireless telecommunication and power transmission? If so, where would be their locations?

Phenomena related to the six sense and ESP are believed to occur solely through the mind without the involvement of any conventional senses. This idea has primarily come into sight from religious doctrines. Many cultures around the world associate paranormal experiences like dreams, visions, clairvoyance, precognition and so on to the concept of the "third eye", an invisible eye that can perceive beyond the ordinary sight. Hindus believe that the third eye is localized in the middle of the forehead just above the junction of the eyebrows. Taoists and many Chinese religious sects share a similar idea. They actually train their third eye by focusing attention on the section between the eyebrows with eyes closed and body positioned in various postures to access the right vibration from the universe. In Taoism, the third eye is regarded as an important energy center of the body. The theosophist H.P. Blavatsky proposed that the third eye is actually the pineal gland of thalamus located between the two hemispheres of the brain.[52] With the shape resembling a pine cone,[53] may the pineal gland really act as human wireless sensing antenna to broadcast and receive information and energy? Well, it is hard to convey but the functions of the pineal gland are still not fully understood, and its deep location within the brain suggests particular importance. Science has so far established that the pineal gland produces melatonin, a hormone responsible for modulating sleep patterns and synchronizing day/night cycles.[54]

On the other hand, the pineal gland is part of the thalamus with important roles in sleep, wakefulness, and consciousness.[55] The thalamus belongs to the limbic system that controls emotion, sensation, and memory. It is physically connected to the hippocampus in charge of controlling episodic, rodent event, recollective, and familiarity memory.[56] Furthermore, the mediodorsal thalamus might contribute to signal strength amplification in the cortex that made our brains capable of making complex decisions.[57] More importantly, the thalamus works as a relay station between sensory pathways and the brain. Except for the olfactory system, every sensory organ has a thalamic nucleus to pick up sensory signals and forward them to the specialized part of the brain, including taste, touch, sight, position, and balance. For example, impulses induced in the eye are sent the thalamus nucleus known as "lateral geniculate", which then project them to the visual cortex of the brain for analysis. The thalamus is also believed to process information as it goes through since each primary sensor relay area gets a feedback connection from the cerebral cortex of the brain. Damage to the thalamus can cause sensory loss.

The thalamus is, therefore, deeply involved in reception, release and even processing of sensory information. However, most mechanisms in control of such processes are still unclear, and one should question whether the thalamus could possibly assist in the remote transmission of information and energy, which might explain paranormal experiences? If so, how? Does the thalamus function as a biological bidirectional antenna for wireless broadcasting and reception of information and energy? Well, this is not impossible since a little is yet known about the sensory tasks of the thalamus. However, from the design standpoint, it is certainly not the best location for an antenna since the thalamus is totally buried inside the human head. Artificial antennas provide the best signal transmission and reception when placed on higher grounds free of physical obstacles for easy propagation of the electromagnetic waves. Accordingly, the greatest locations for biological antennas should be the most exposed parts of the human body.

One obvious possibility of such location is simply hair that covers many parts of the human body. Hair grows following cycles comprised of three distinct and concurrent phases that determine its length. Humans have different types of hair, including vellus hair and androgenic hair. The vellus hair develops on most of the human body during childhood, starting from 36 to 40 weeks of gestation.[58] Androgenic hair, also known as terminal hair, grows on the human body during and after puberty. Hair grows from a pouch, called "follicle" found in the dermis layer of the skin. Inside follicles, there are sebaceous glands responsible for producing oil or sebum, which give hair its soft, shiny, and sometimes oily aspect. The apocrine gland engenders sweat and opens into the follicle. Around the follicle, there is the arrector pili muscle that controls the physical movement of hair. Humans have follicles all over their skins, except some glabrous and bald parts that closely interact with the surrounding, such as lips, palms, feet soles, labia minora for women, and glans penis for men.[59] The sensors and receptors connected to hair are situated in the follicles and inside the skin. Movement of hair due to air flow or physical contact is sensed by receptors around the hair strand. The induced impulses are then sent to the brain *via* the thalamus for examination.[60] Some hair strands are more sensitive than others to protect our vital organs, such as eyelashes and eyebrows.[61] For instance, eyelashes will close upon sensing of air particles or dirt to prevent them from getting inside the eyeball. Some mammals have longer, thicker hairs as tactile sensors in the form of eyelashes, eyebrows, whiskers or furs. Signals from these hairs are interpreted by the neocortex of the brain to provide a better view of the surrounding, which may have helped in the survival of these animals. [62]

Therefore, hair is obviously involved in sensing, yet the mechanisms are still not fully compressible. Throughout history, the purpose of hair has always been a subject of debate and remains shrouded in mystery even today. Many theorize that hair serves for protection against dust, sweat, and sebum, as well as regulation of the body temperature. But is this all when it comes to hair functions? For instance, why vellus hair develops during gestation and androgenic hair grows during and after puberty? It is possible that vellus hair assists our senses developed before puberty as wireless components, and androgenic hair aids in sexual stimulation? In other words, is hair our biological antennas that help us to remotely detect changes in stimuli?

From the composition and structure viewpoints, hair is actually a fiber composed of 91% of a hard, compact and strong protein called keratin that strongly bonds together. The hair shape looks like common type low gain omnidirectional antennas applied in radio communication devices. Both hair and artificial antennas share similar behaviors too. During weather storms, omnidirectional antennas often channel lightning strike induced-electric charge that travels through them. Likewise, if a person gets struck by lightning, its hair will stand on end before being hit by the strike. A famous story of a close encounter with lightning came from a photograph taken in 1975, of two young brothers on a family vacation in Sequoia National Park, California. The photograph depicts the boys apparently amused by the effect of the static electricity on their hair, making it stick straight up and outwards, unaware of the danger waiting to happen. One of the boys was struck by lightning and suffered third-degree burns, just minutes after the photograph was taken. However, unlike artificial antennas, hair is not an electrical conductor but an insulator. It conductivity might slightly increase under wet conditions due to its enhanced proton conductivity. But, is conductivity really important in broadcasting and reception of electromagnetic waves?

Not necessarily. For instance, an optical fiber is a medium that transmits information coded in a beam of light instead of an electric current in conventional metal conductors. Optical fibers are actually electric insulators, generally made by drawing glass or plastic to diameters slightly thicker than those of the human hair.[63] Despite this, they carry substantial amounts of data over long distances with lower attenuation and no interference.[64,65] In the past, optical fibers were employed in endoscopy to help doctors view inside the human body without surgery. Today, they are widely utilized as media for long distance transmission of information and power. The small size of optical fibers made them valuable for remote sensing to monitor changes in temperature, pressure, strain, among other quantities. Common cases comprise extrinsic fiber optic sensors for inaccessible places, such as inside aircraft jet engines and electrical transformers. Fiber optic sensors come also handy in advanced intrusion detection security systems and as components of optical biosensors and optical chemical sensors.[66] Moreover, optical fibers are employed in photovoltaic cells to convert light into electrical power,[67] as well as in high-voltage transmission equipment to power electronics in high-powered antennas and computing devices. Optical fibers also find applications in antenna systems designed for wireless communication with extended radio frequency coverage in places hard to penetrate, such as inside large building and airports.

Thus, it is not unreasonable that hair might act as fiber biological antennas for remote transmission of information and power. That is to say, every human identified sense comprises a wireless component for remote monitoring of stimuli. In addition to androgenic hair that may well serve in remote sexual stimulation for mating purposes, vellus hair is present around all traditional senses, including the noise cavity and ceiling, above the lips, inside the ears, eyelashes and eyebrows, and most parts of the skin, especially on top of each finger and toe. But what may be the real purpose of such hair found at these specific places?

There is an old saying that itches in different body parts held meaning. For example, an itch on the right or left-hand palm signifies that you are going to receive or lose money. If the right or left eyebrow itches, someone is saying good or bad things about you. An itchy lip indicates that someone is going to kiss you, right or left depending on how close this person to the family. Left or right ear ringing, you are about to hear good or bad news. Itchy-foot, you will go on a long journey. Eye moving means that you are about to meet someone that you did not see in a long time, right or left depending on how close this person to the family. May these old superstitions possibly hold some truth or they are simply self-delusions?

Well, if hair serves as biological antennas embedded within each known sense to broadcast and receive information remotely, then these superstitions may be justified. If so, can these be interpreted as past, present or future events? Technically, most artificial wireless sensors monitor and detect real-time changes in stimuli, unless the sensor is programmed to simulate future events built on previous data. Sometimes, information passed on by wireless sensors can be delayed due to weather or other interferences. Similarly, if the hair is a biological antenna that transmits information from one person to another, this would likely be real-time events, unless these senses are so advanced that they can even simulate future experiences based on previously memorized information. For instance, an itch in the right-hand palm announcing coming money can be a real-time response to someone thinking or planning to hand you money. During this course, the provider of money broadcasts his thoughts *via* its biological antennas. After modulation of the thoughts, the impulses come out as electromagnetic waves, just like in artificial antennas. The information then journeys through space before hitting the biological antennas of the receiver of money. At that time, the collected electromagnetic waves are then demodulated back into impulses, which after transiting through the thalamus and specialized parts of the brain are translated into an itchy hand palm. Since the money is often received at later dates after the initial planning and thinking, the process may look like a prediction of future events but actually is just real-time detection.

Whether the human wireless senses are located in the thalamus, hair or in other body areas, the question is why most paranormal events can not scientifically be replicated nor verified? The answer might simply lie in the efficiency of the human wireless senses. For instance, not all artificial antennas perform in the same way but their efficiencies depend on several parameters. One important factor that determines the efficiency of an antenna is the radiation pattern, representing the area of space covered by the signals transmitted by the antenna. Log-periodic antennas operate in a wide range of frequencies whereas small loop antennas work in a narrow range of frequencies. Beyond this range, the signal strength reduces and transmission becomes inefficient. There are two types of antennas. High gain antennas focus most of their energies in a particular direction, covering longer range with better signal quality. However, both the transmitting and receiving antennas must carefully be positioned for improved connection. A common model of this type of antennas is the parabolic dish utilized as satellite television signal receiver. By comparison, low-gain antennas focus their energies over a wider range but the position does not really matter. Models of this kind of antennas are those used in portable radios and cordless phones. The efficiency of a transmitting antenna is assessed by the ratio of energy broadcasted in all directions to the energy absorbed by the antenna terminals. Excess energy is converted into heat due to the low resistance of the antenna conductors or to the ineffective communication between the antenna's components. The physical structure of the antenna and its position relative to the earth's surface also count in the determination of its efficiency. A simple straight wired antenna will have one polarization when positioned vertically and another polarization when placed horizontally. For an antenna placed on the ground, its height relative to the ground and the ability of the ground to pass electric charge might impact its efficiency.

In a similar manner, the efficiency of the human biological antennas in broadcasting and reception of information and energy would depend on many factors, encompassing the physical and moral status of the person, its location, the weather and so on. Because we are dealing with biological senses, evaluating the impact of each factor can be of extreme complexity, and it is not clear from the literature or gathered personal accounts whether these factors boost or interfere with signal transmission. For instance, if the person is sick, its internal senses try to monitor the illness, which might interfere with the wireless signal reception. Still, the pain engendered by the disease may broadcast stronger signals that will be picked up by other peoples that the sick person thought about. Likewise, many accounts claim that telekinesis is triggered by intense emotional states, and often manifested as an electric discharge at the receiver's end that might be located far-off. Occasionally, for unclear reasons, the electric discharge does not grasp the receiver and instead becomes lost in the atmosphere, contributing to weather phenomena. Other physical and moral states like hunger, rest, exhaustion, anger, happiness, pain and so on may also impede or strengthen the signal transmission. The nature of the ingested food and its metabolism also count. Several accounts suggest that the location plays a significant role in signal reception and broadcast. A person sleeping in a room built of insulating materials for penetration of electromagnetic waves will unlikely pick up or broadcast any signals. By contrast, exposed areas permit better communication and energy transmission. The geographical location also matters due to the earth's geomagnetism. A number of personal accounts stated noticeable variation in dreaming states as a function of the site on Earth. This may explain why ancient peoples like Egyptians and Greeks slept in sanctuaries and temples to help them receive more vivid dreams. If the hair is a human biological antenna, its condition like length, cleanness, humidity, temperature, and so on might affect the efficiency of signal transmission. If so, could this be the reason why the religions Sikhs have an obligation not to cut their hair as they view this as a form of apostate, meaning fallen from the religion? Can their long hair possibly help in strengthening the signal reception and assist them in their quest of spiritual enlightenment?[68]

If the human wireless senses in fact exist, one has to wonder how is it possible to grasp information about the past, present, and future *via* dreams and telepathy. Are these senses supposed to deliver information in real time? Well, not all broadcasted information is instantly received but occasionally interferences cause a delay in data transmission. For instance, you may have noticed that when you check your Wi-Fi surveillance cameras on your smartphone, you sometimes download videos that have been sent second or even hours ago instead of a real-time recording. But why? A very important part of our technology today consists of data storage media, which make our personal computers, tablets, mobile telephones, and digital media work as they should. A digital storage medium is a piece of hardware employed to store data when they are not in use, and retrieve the said data when need to be accessed. Common data which can be stored comprise documents, images, audio, video, and software. There are two categories of digital memory storage. The first is the non-volatile form, which stores the information on the device and can be retrieved later even after turning the device off and on again. This type of storage media is utilized in personal computers and laptops, mobile phones, DVDs, CDs, cassette tapes and so on. Also, it is this type of memory storage that made it possible to view the delayed videos on your smartphone. Because your NVR is equipped with a hard disk memory storage, all recorded videos were stored in the disk and can be accessed remotely. In other words, without the hard disk, you can only see real-time videos and nothing from the past. The second type of memory storage media is volatile, which must continuously be powered to keep the data. That is to say, if the power supply is interrupted or the device is turned off, the data is instantly lost. This type of storage media is particularly used in banking systems to protect sensitive information.

In our time, memory storage media can not only be utilized locally but might also be accessed remotely. You most likely heard about "cloud computing", which uses a network of computers connected through the internet to store, process, and manage digital data wirelessly. Today, many popular companies, such as Google *via* Google Drive and Microsoft *via* OneDrive offer cloud storage services where users might store information, such as documents, photos, videos and music in their cloud networks. This trend is becoming increasingly common since cloud computing data are accessible worldwide through the internet, which resolves problems associated with transportation and theft. But how does cloud computing actually operates? First, cloud computing companies set up large facilities of computers equipped with large capacity memory storage media. These computers are then interconnected through the internet to create large networks or virtual space otherwise known as the "cloud", where the digital data can be kept and retrieved from anywhere within the network. Users are provided with unique identifiers and passwords and once logged, they can access the resources existing from all computers in the network. Some companies like IBM have even added quantum computers to their cloud networks for users to crack mathematical problems. Keep in mind that most technologies that made up the cloud, including the computers (laptops) and Wi-Fi routers have antennas for remote signal transmission.

For that reason, if humans really own biological antennas, does this mean that information broadcasted by a person *via* its wireless senses is stored somewhere and can be accessed later by another person through dreams and telepathy, just like in cloud computing? Well, in retrocognitive or superconscious telepathies, a person could access information from the past or tap into universal knowledge. This likely suggests the existence of some sort of information storage media. In artificial storage media, especially the non-volatile forms, digital data can be saved in many types of materials. The most useful are semiconductor-based integrated circuits utilized to stock data in the form of one bit of binary information 1 and 0.[69]

Until altered through set/reset processing, each coded bit of information is stored and can be accessed and read. This storage technology based on binary code might also be built from magnetic materials, such as ferrite cores or magnetic bubbles.[70] Alternatively, some materials like polymers are employed for data storage as an electric charge.[71] More fascinating, researchers even succeeded to keep data in DNA nucleotides with a capacity up to 1.28 petabytes per gram of regular DNA or 215 petabytes per gram of algorithms DNA known as "DNA fountain".[72]

So, information storage media exist all around us in one form or another, starting by our DNA that may actually store our thoughts, actions, emotions, and so on. This information can later be retrieved by another person thru dreams or telepathy even after passing away since bones contain DNA that can stock coded data. Yet, DNA memory storage will hardly explain some retrocognitive telepathy cases and most superconscious telepathy. A person reading the past of an object simply by touching it signifies that the information is encoded or stored within the object itself or someplace but unlikely in DNA. Similarly, if somebody could tap into the universal knowledge by means of superconscious telepathy, this implies that the data is stocked beyond the human body and possibly on Earth or in the outer space.

Material with data storage properties, such as semiconductors and magnetic materials are present on both our planet and in the outer space. One form is simply rock or stone, which covers the Earth's crust. The rock contains minerals (such as silica composed of silicon and oxygen), metals (such as iron, cobalt, and nickel), organics, refractories and other resources to fashion the best semiconductor, magnetic or any form of memory storage media.[73] In the outer space, many of celestial objects nearby our planet might also store information. For instance, millions of asteroids travel nearby Earth, especially in our inner solar system and in the asteroid belt. Large asteroids contain rocky core enclosed in icy mantles,[74] and may hold traces of amino acids and organics. As a result, just like rock on our planet, asteroids might store information.

Solid debris from asteroids forms meteorites that occasionally survive their passage through the Earth's atmosphere to reach the surface without burning out. Ancient peoples believed that some meteorites possess mystical powers and they are messages from the gods or direct communication with the universe. For example, the temple Apollo in ancient Greece, which is home for the famous oracle of Apollo, was built at the place where a meteorite fell from the sky. The Willamette meteorite, which is the largest rock ever fell from the sky in North America, is considered by the American native Indians to grasp special powers. But, it is conceivable that what ancient peoples believed is simply information or power stored and encoded within these stones, and retrieved by some through their strong wireless senses? If so, what kind of information or energy is really coded within these stones? Is it that broadcasted by ancient humans or from other galactic civilizations (if any)? If this is true, are our advances in science and technology come as a result of hard work or some may have originated by tapping into information preserved in materials around us and/or in our universe?

In theosophy and anthroposophy, such notion exists and has been described as the "Akashic records", which include all the human-related events, such as thoughts, emotions, and words occurred in the past, present and future. This information encoded in a non-physical plane of existence called the "etheric plane",[75] has so far not yet scientifically confirmed. But is the proposed concept of "broadcast-storage-retrieval" of information and energy

44

through the human wireless senses a good match of such records? Of course, "broadcast-storage-retrieval" cannot explain future events. However, it is feasible that by tapping into such records, one may learn lessons from the past to make sufficiently accurate predictions to impact the future of mankind.

If this is correct, then many other paranormal phenomena and beliefs could be clarified through the suggested concepts of "broadcast-reception" and "broadcast-storage-retrieval" of information and energy for present and past events, respectively. Reincarnation, for instance, is an old idea that is carried across time into the modern world. Although its definition varies from one religion to another, reincarnation is mostly described as the transfer of the soul of a dead person into another body at the time of its birth, known as rebirth or transmigration.[76] This idea of reincarnation has originated from ancient Indian religions, namely Hinduism, Jainism, Sikhism, and Buddhism.[77] Each Indian religion has a series of doctrines called "Samsara", summarizing the cycles of life. For example, the ultimate goal of a Buddhist is to break the cycle of reincarnation, leading to the liberation of the soul, also known as "Nirvana".[78] Since it is nearly impossible to achieve Nirvana without sufficient wisdom and enlightenment, most Buddhists are focused on living ethically by doing meditation and other yogic practices that help them accumulate good karma, which determines the form they will take in the next life.[79] Nirvana is easier to achieve by higher level beings. By comparison, earthly beings should have a hard time achieving Nirvana than heavenly beings but easier than animals or lower forms. In Judaism, humans only reincarnate as other humans and at the same sex. Many other religions and belief systems, both new and old, preach similar concepts of reincarnation. The Druze peoples, an Arabic-speaking western-Asian ethnoreligious group, believe in the transmigration of the soul and end of the cycle of reincarnation by unifying with the great Cosmic Mind, similar to the Buddhist concept of Nirvana. Many philosophers, such as Pythagoras, Socrates, and Plato all held a common belief in reincarnation that they called "metempsychosis", meaning transmigration.[80]

But is this notion of reincarnation merely the body's ability to broadcast and retrieve stored information? Using human wireless senses, we basically send out information about our remarkable events or strong thoughts during our lifetimes. This process would likely peak at the time of death since most senses, including the wireless components, become highly activated due to unbearable pain engendered by illness, incident, homicide, emotional distress and so on. Accordingly, enormous amounts of data are wirelessly broadcast at the time of death and become stored in materials around the deceased person or in the outer space. Newborn children can access the kept data *via* their wireless senses. The efficiency of information retrieval relies on the strength of their biological antennas, physical and mental status, location, and many other factors. Let's say the information is stored in a moving asteroid during its passage around the Earth's atmosphere, children close to that asteroid should have better access to the data, just like a telecommunication satellite for cell phone or GPS signal reception. The retrieval of information by these newborn children do not happen instantly at the time of birth but rather progressively as they grew learning step by step. Because not all information of the deceased person is broadcasted to form a full personality, some children of this group retrieve part of the stored information and continue on the footstep of the deceased person by diffusing out their own data. This broadcast-storage-retrieval process of information carries on for many death and rebirth cycles, and bit by bit, full personality forms. This corresponds to the end of reincarnation and liberation of the soul, creating a "spirit", which can be good or bad

47

depending on the content of the coded information. One way to comprehend this process is to think of it as a computer program developed by scientists through cloud computing. During the first year, some scientists initiate the program by putting some data on the cloud. In the second year, other scientists retrieve the information, improve it and put it back into the cloud. Year after year, more data are stored in the cloud and the program advances until completion of a functional version, which can then be liberated for marketing or other use. Just like spirits, the computer program can be good or bad, such as a novel cell phone application or a malicious virus with destructive intents.

On the other hand, is it conceivable according to the proposed concept of broadcast-storage-retrieval of information that humans reincarnate into lower forms, such as animals or insects? The answer is unlikely since animals and other lower forms do not make use of the human language, which will prevent them from retrieving the data in an understandable manner. For instance, cloud computing digital data written in a certain programming language (such as FORTRAN) would doubtfully be read it in another programing language (such as Ada or Pascal). However, both man and women can equally access the stored information, meaning that man may reincarnate as women, and vice versa. However, due to the difference in personality between man and women, newborn children would be more driven to retrieve stored information that better fit their gender. As a result, man will mostly reincarnate as man and women as women. Occasionally, newborn children may tap into stored information of the opposite sex. This engenders unconventional behaviors, such as a man acting like a women, and vice versa. Whether animals and lower forms can reincarnate into their own species, this will depend on the efficiency of their wireless senses. If the hair is biological antennas, then animals should have more advanced wireless senses than humans. Still, information processing by the thalamus and brain of animals is rudimentary when compared to humans. Hence, animals ought to better comply with broadcast-reception of information in real time to remotely detect threats and suitable mates that help in survival. However, the broadcast-storage-retrieval process of information that grants reincarnation is undeveloped in animals and lowers forms due to lack of appropriate thought processing. The seventeenth-century philosopher and scientist René Descartes regarded the pineal gland of the thalamus as the place of the soul at which all thoughts are formed.[81]

Other paranormal experiences can as well be interpreted *via* broadcast-storage-retrieval of information through the human wireless senses. We all heard of haunted houses often linked to tragic events, such as homicide, accidental death, or suicide. During horrific death, the wireless senses of the deceased person become so activated that they broadcast most of the information associated with the crime. The data are then stored in materials like stone available in the area of the crime and can be accessed later by other folks who have efficient wireless senses. Similarly, are supernatural experiences, such as prophecy, revelation and divination, only the result of broadcast-storage-retrieval of information? That is to say, did ancient prophets directly tap into stored knowledge, broadcasted by ancient humans or possibly others civilizations from the outer space, and then mistaken for a divine word? Well, it is not a hypothesis to rule out. As said, the stored information may well improve after several broadcast-storage-retrieval processes corresponding to death-rebirth cycles. The final encoded information would look advanced just like a finished computer program available in the cloud computing that passed by several stages of development and improvement. Individuals with powerful wireless senses, such as prophets, should then be able to retrieve such data. If this is true, are the ten stones tablets unearthed by Moses on the Mount Sinai the result of remote sensing? This may seem odd but the possibility is there. Keep in mind that our development in science and technology comes as a result of mimicking what nature already fashioned. Still, our advances look rudimentary when compared to those accomplished by nature's laboratory. Consequently, it is not impossible that the stones tablets holding the Ten Commandments are actually advanced materials able of storing the right encoded information, which was accessible to Moses through his wireless senses. Otherwise, why Moses was obliged to move to that particular place, which is the top of Mount Sinai, to retrieve such information? And, this is only one case among many others, where prophets necessitate moving to a particular place to receive the divine word. Think of this as signal reception by cell phones or GPS devices. These days, the signal reception improved remarkably but there was a time when we were required to move around with the cell phone in hand looking for a better area with decent signal reception.

Most examined paranormal occurrences dealt so far with broadcast-storage-retrieval of information but can this process also explain experiences involving the manifestation of energy, such as curses and spirits? Well, like information storage, materials with the faculty to stock power also exist and have been the subject of intense research and development for centuries. These materials integrated within devices, known as accumulators or batteries, serve as media to retain the power and store it for later use. Different forms of energy, such as chemical, electrical, thermal or heat, radiation, and gravitational potential may be captured and stored for later usage. A common example is the lithium-ion batteries utilized to power our cell phones. Typically, this type of batteries is composed of materials like carbon and metal oxides,[82] which are abundantly present in nature. Meanwhile, the types of materials and their arrangements are crucial in successfully advanced energy storage devices. Yet, we must recall that before such materials were explored in artificial laboratories, they readily existed and have been the subject of experimentation by nature's laboratory for billions of years. Therefore, advanced materials for energy storage should naturally exist on our planet and/or throughout the universe, making it feasible to store energy *via* the broadcast-storage-retrieval concept. If so, how?

The human body works like a biofuel cell to convert consumed and assimilated food into mainly thermal and electrical power. Depending on many circumstances that remain unclear, some of this energy is broadcasted through the human wireless senses. In telekinesis, the energy transfer takes place in real time according to the concept of broadcast-reception, and the effects are immediately noticeable on the receiver's end. Yet, in most cases, this energy becomes trapped and stored in materials, then manifest itself later through individuals with efficient human senses. One popular account of such paranormal phenomena is the "Curse of the Pharaohs" that supposedly killed several members of the Howard Carter's team who excavated the tomb of Tutankhamun (KV62) in 1922. While this might be explained by the presence of bacteria and fungi in the Tutankhamun's tomb, the proposed concept of broadcast-storage-retrieval of energy can also provide an alternate interpretation. During mummification of a pharaoh, the priests spend an incredible time in preparing the corpse to journey into the afterlife. The pharaoh Tutankhamun died unexpectedly young at the age of 18.[83] This might profoundly sadden the priests, which highly activated their wireless senses and triggered telekinetic episodes. Hence, significant amounts of energy related to the event have been broadcasted, and mostly trapped in the tomb itself and in the surrounding made of stone like limestone, granite, mud break, among others. When Howard Carter and his team opened the tomb, they knew that entering the tomb is a form of trespassing, which triggered their wireless senses for retrieval of the stored energy.

In sum, YES, our current science may provide some answers to paranormal phenomena. The goal of this essay was not to enfeeble the general public that oracles and fortune tellers are real but rather to question the credibility of a few documented and personal accounts that may hold some truth. The objective was to quest for rational reasons according to our current advances in sciences and technology. I personally believe that the human wireless senses exist but whether they are the reason behind paranormal phenomena remains open for debate. These wireless senses are our advanced biological tools for remote diffusion and pick up of information and energy coded in beams of electromagnetic waves. However, how such processes occur and the mechanisms that trigger them are still unclear. These senses can prevail under certain circumstances and fail under others. The precise locations of such senses are still shrouded in mystery but two oppositions were provided. The first being the thalamus and the second was simply hair as biological antennas. Using the human wireless senses, paranormal experiences like dreams, telepathy, and telekinesis may be explained through real-time broadcast-reception of information and energy from one human to another. Occasionally, the transfer of information and energy does not reach the receiver in real time but instead, become trapped and encoded in storage media made of materials around us or in the universe. This gives rise to the concept of broadcast-storage-retrieval of information and energy, which might clarify paranormal occurrences like reincarnation, divination, revelation, prophecy, curses, among others. The identification of such senses would surely open new avenues for practical transmission of communication and energy. Once reinforced technologically using implants, future communication, and energy transfer will hopefully take place telepathically and telekinetically *via* the human wireless senses instead of cell phones, email messaging, remote controls, among other gadgets.

References

(1) Black J., Green A. (1992). Gods, Demons and Symbols of Ancient Mesopotamia: An Illustrated Dictionary. Austin,

Texas: University of Texas Press (pp. 71–72), 89–90.; Oppenheim, L. A. (1966). Mantic Dreams in the Ancient Near East in G. E. Von Grunebaum & R. Caillois (Eds.), The Dream and Human Societies (pp. 341–350). London, England: Cambridge University Press.

(2) Lincoln J. S. (1935). The dream in primitive cultures London: Cressett.

(3) Bulkeley K. (2008). Dreaming in the world's religions: A comparative history (pp. 71–73).

(4) Krishnananda S. (16 November 1996). "The Mandukya Upanishad, Section 4".

(5) Tedlock B. (1981). "Quiche Maya dream Interpretation". Ethos. 9, 4, 313–350.

(6) Jung, 1964 (p. 21).

(7) Carroll R. T. (2005). "The Skeptic's Dictionary, Telepathy". Skepdic.com.

(8) Eshel O. (December 2006). "Where are you, my beloved?: On absence, loss, and the enigma of telepathic dreams". The International Journal of Psychoanalysis. 87. 6, 1603–1627.

(9) Ullman M. (2003). "Dream telepathy: experimental and clinical findings". In Totton, Nick (ed.). Psychoanalysis and the paranormal: lands of darkness. Reference, Information

and Interdisciplinary Subjects Series. Karnac Books (pp. 14–46).

(10) Simon N. (2006). The Last Explorer: Hubert Wilkins, Hero of the Great Age of Polar Exploration. Arcade Publishing (pp. 267-268).

(11) Glossary of Parapsychological terms-Telepathy (2006-09-27), Wayback Machine - Parapsychological Association.

(12) Muldoon S. (2007). Psychic Experiences of Famous People (Reprint ed.). Kessinger Publishing (pp. 55–56).; "Online Etymology Dictionary". (August 28, 2011). Telekinesis. 1890, said to have been coined by Alexander N. Aksakof (1832-1903) Imperial Councilor to the Czar... Translates Ger. 'Fernwirkung.'

(13) Psychokinesis (1914).... "Parapsychology Foundation "Basic terms in Parapsychology"". (August 28, 2011).; Holt, H. (1914). On the Cosmic Relations (PDF). Cambridge, Massachusetts, USA: Houghton Mifflin Company/Riverside Press.

(14) Wiseman R. (1997). Deception & Self-deception: Investigating Psychics. Amherst, New York: Prometheus Books (pp. 182–196).

(15) Green E., Green A. (1977). Beyond Biofeedback(2nd ed.). New York: Delacorte Press/S. Lawrence (pp. 197–218).

(16) Berger A. S., Berger J. (1991). The Encyclopedia of Parapsychology and Psychical Research (1st ed.). New York: Paragon House (pp. 326), 341, 430.; Paraphysics R&D - Warsaw Pact (U). Prepared by U.S. Air Force, Air Force Systems Command Foreign Technology Division.

(17) Hall J. (2011). Guyton and Hall Textbook of Medical Physiology (12th ed.). Philadelphia, PA: Saunders/Elsevier.

(18) The Importance of the Sense of Touch in Virtual and Real Environments" (PDF). International Society for Haptics.

(19) Wu L. Q., Dickman J. D. (May 2012). "Neural correlates of a magnetic sense". Science. 336, 6084, 1054–1057.

(20) Cattle shown to align north-south". BBC NEWS - Science/Nature.

(21) Dunn B. D., Galton H. C., Morgan R., Evans D., Oliver C., Meyer M., Cusack R., Lawrence A. D., Dalgleish T. (December 2010). "Listening to your heart. How interoception shapes emotion experience and intuitive decision making". Psychological Science. 21, 12, 1835–1844.; Shah P., Hall R., Catmur C., Bird G. (August

2016). "Alexithymia, not autism, is associated with impaired interoception". Cortex; A Journal Devoted to the Study of the Nervous System and Behavior. 81, 215–220.

(22) Farr O. M., Li C. S., Mantzoros C. S. (May 2016). "Central nervous system regulation of eating: Insights from human brain imaging". Metabolism. 65, 5, 699-713.

(23) Kleckner I. R., Wormwood J. B., Simmons W. K., Barrett L. F., Quigley K. S. (November 2015). "Methodological recommendations for a heartbeat detection-based measure of interoceptive sensitivity". Psychophysiology. 52, 11, 1432–1440.

(24) "Brain Areas Critical To Human Time Sense Identified". UniSci – Daily University Science News (2001).

(25) Kinnavane L., Amin E., Olarte-Sánchez C. M., Aggleton J. P. (November 2016). "Detecting and discriminating novel objects: The impact of perirhinal cortex disconnection on hippocampal activity patterns". Hippocampus. 26, 11, 1393–1413.

(26) Metzinger, T. (2003). Being No One (p. 508).

(27) Wegner D. M., Wheatley T. (July 1999). "Apparent mental causation. Sources of the experience of will". The American Psychologist, 54 (7), 480-492.

(28) Parker K. L., Lamichhane D., Caetano M. S., Narayanan N. S. (October 2013). "Executive dysfunction in Parkinson's disease and timing deficits". Frontiers in Integrative Neuroscience. 7, 75.

(29) Ivancevic, V., Tijana (2005), Natural Biodynamics, World Scientific (p. 602).

(30) Atema, J. (1980) "Chemical senses, chemical signals, and feeding behavior in fishes" (p. 57–101). In: Bardach, JE Fish behavior and its use in the capture and culture of fishes', The WorldFish Center.

(31) The illustrated story of the Vampire bat".

(32) Marshall J., Oberwinkler J. (October 1999). "The colourful world of the mantis shrimp". Nature. 401, 6756, 873–874.; Octopus vision, it's in the eye (or skin) of the beholder.; Study proposes explanation for how cephalopods see color, despite black and white vision.; Odd pupils let 'colorblind' octopuses see colors.

(33) Carl Z., "The Search for Genes Leads to Unexpected Places", New York Times (April 27, 2010) (page D1), New York edition Plants sensing gravity.

(34) "Electroreceptive Mechanisms in the Platypus" (1999-02-09).; Drake N. (2011). "Life: Dolphin can sense electric fields: Ability may help species track prey in murky waters". Science News. 180, 5, 12.

(35) Bullock T. H. (2005), Electroreception, Springer (pp. 5–7).; Morris, S. C. (2003), Life's Solution: Inevitable Humans in a Lonely Universe, Cambridge University Press (pp. 182–85).

(36) Kandel E., Schwartz J., Jessell T. (2000), Principles of Neural Science, McGraw-Hill Professional (pp. 27–28)

(37) Tleis N. (2008), Power System Modelling and Fault Analysis, Elsevier (pp. 552–554).

(38) Grimnes S. (2000), Bioimpedance and Bioelectricity Basic, Academic Press (pp. 301–309).

(39) Sears F. et al. (1982), University Physics, Sixth Edition, Addison Wesley.

(40) Purcell and Morin, Harvard University. (2013). Electricity and Magnetism (820p) (3rd ed.). Cambridge University Press, New York(p 430).

(41) Maxwell J. C. (1 January 1865). "A Dynamical Theory of the Electromagnetic Field". Philosophical Transactions of the Royal Society of London. 155, 459–512.

(42) "Wired Communications". The Great Soviet Encyclopedia (3rd ed.). The Gale Group, Inc. 1979.

(43) "What is wireless communication technology and its types". Engineers Garage (22 June 2017).

(44) "ATIS Telecom Glossary 2007". atis.org. (2008)

(45) "Radio Frequency, RF, Technology and Design, Radio Receiver Technology". Radio-Electronics.com.(27 January 2012).

(46) Graf, R. F. (1999). Modern Dictionary of Electronics. Newnes (p. 29).

(47) Balanis C. A., Ioannides P. I. (2007). Introduction to Smart Antennas. Morgan and Claypool. (p. 22).; National Telecommunication Information Administration (1997). Federal Standard 1037C: Glossary of Telecommunications Terms. US General Services Administration (pp. 0-3).

(48) Chapman R.F. (1998). The Insects: Structure and Function (PDF) (4th ed.). Cambridge University Press (pp. 8–11).; Boxshall G., Jaume D. (2013). Functional Morphology and Diversity: Antennules and Antennae in the Crustacea. Oxford University Press (pp. 199–236).

(49) Merlin C., Gegear R. J., Reppert S. M. (September 2009). "Antennal circadian clocks coordinate sun compass orientation in migratory monarch butterflies". Science. 325, 5948, 1700–1704.

(50) Ellis P., Brimacombe L. (1980). "The mating behaviour of the egyptian cotton leafworm moth, Spodoptera littoralis (Boisd.)". Animal Behaviour. 28, 4, 1239–1248.

(51) Justus K. A., Mitchell B. K. (November 1996). "Oviposition site selection by the diamondback moth, Plutella xylostella (L.) (Lepidoptera: Plutellidae)". Journal of Insect Behavior. 9, 6, 887–898.; Scriber J. M. (Michigan State University, East Lansing, R.V. Dowell (1991). "Host plant suitability and a test of the feeding specialization hypothesis using Papilio cresphontes (Lepidoptera: Papilionidae)". The Great Lakes Entomologist (USA).

(52) Macchi M. M., Bruce J. N. (2004). "Human pineal physiology and functional significance of melatonin". Front Neuroendocrinol., 25, 3–4, 177-195.; Arendt J., Skene D. J. (2005). "Melatonin as a chronobiotic". Sleep Med Rev., 9, 1, 25–39.

(53) "Pineal (as an adjective)". Online Etymology Dictionary, Douglas Harper (2018).

(54) Motta M., Fraschini F., Martini L. (1967). "Endocrine Effects of Pineal Gland and of Melatonin". Exp Biol Med (Maywood). 126, 2, 431–435.; Naidoo V., Naidoo S., Mahabeer R., Raidoo D. M. (2004). "Cellular distribution of the endothelin system in the human brain". Journal of Chemical Neuroanatomy. 27, 2, 87–98.

(55) Gorvett, Z. "What you can learn from Einstein's quirky habits". bbc.com.

(56) Aggleton J. P., O'Mara, S. M., Vann, S. D., Wright N. F., Tsanov, M., Erichsen, J. T. (2010). "Hippocampal-anterior thalamic pathways for memory: Uncovering a network of direct and indirect actions". European Journal of Neuroscience. 31, 12, 2292–307.; Burgess N., Maguire E. A, O'Keefe J. (2002). "The Human Hippocampus and Spatial and Episodic Memory". Neuron. 35, 4, 625–641.

(57) New Role Discovered For Brain Region". Neuroscience News (2017).

(58) Heather B. "Vellus Hair - Peach Fuzz & Puberty Hair Growth". About.com.

(59) Prescott T. J., Ahissar E., Izhikevich E., eds. (2016). Scholarpedia of Touch. San Diego, USA: Atlantis Press. (p. 9).; Linden D. J. (March 2015). "Chapter 2". Touch: The Science of Hand, Heart and Mind. Viking.

(60) Dean I., Siva-Jothy M. T. (2011). "Human fine body hair enhances ectoparasite detection". Biology Letters. 8, 3, 358–361.

(61) "Neuroscience for Kids–Receptors". Faculty. washington.edu. (18 February 2015).; "hair biology–functions of the hair fiber and hair follicle". Keratin.com. (18 February 2015).; Sabah N. H. (1974). "Controlled stimulation of hair follicle receptors". Journal of Applied Physiology. 36, 2, 256–7.25.; Montagna W. (1985). "The

evolution of human skin (?)". Journal of Human Evolution. 14, 3–22.

(62) Rowe T. B., Macrini T. E., Luo Z. X. (19 May 2011). "Fossil Evidence on Origin of the Mammalian Brain". Science. 332, 6032, 955-957.

(63) Optical Fiber". www.thefoa.org. The Fiber Optic Association (2015).

(64) Bozinovic N., Yue Y., Ren Y., Tur, M., Kristensen P., Huang H., Willner A. E., Ramachandran S. (2013). "Terabit-Scale Orbital Angular Momentum Mode Division Multiplexing in Fibers". Science. 340, 6140, 1545–1548.

(65) Senior J. M., Jamro M. Y. (2009). Optical fiber communications: principles and practice. Pearson Education (pp. 7–9).

(66) Bănică, F-G (2012). Chemical Sensors and Biosensors: Fundamentals and Applications. Chichester: John Wiley and Sons (Ch. 18–20).

(67) IEEE Spectrum: Electricity Over Glass. IEEE Spectrum (October 2005).

(68) Dilgeer, H. S. (2005) Dictionary of Sikh Philosophy, Sikh University Press.

(69) Fletcher W. (1980). An engineering approach to digital design. Prentice-Hall (p. 283).

(70)   Denny D. T., Yuan-Jen L. (2010). Magnetic Memory: Fundamentals   and   Technology. Cambridge University Press (p. 91).

(71) New Method Of Self-assembling Nanoscale Elements Could Transform Data Storage Industry Archived 1 March 2009 at the Wayback Machine. Sciencedaily.com (2009).

(72)   Yong Ed. "This Speck of DNA Contains a Movie, a Computer Virus, and an Amazon Gift Card". The Atlantic (2017).; "Researchers store computer operating system and short movie on DNA". Phys.org. (2 March 2017).; "DNA could store all of the world's data in one room". Science Magazine (2 March 2017).; Erlich Y., Zielinski D. (2 March 2017). "DNA Fountain enables a robust and efficient storage architecture". Science. 355, 6328, 950-954.

(73) Cipriani N. (1996). The encyclopedia of rocks and minerals. New York: Barnes & Noble.

(74) Asteroid or Mini-Planet? Hubble Maps the Ancient Surface of Vesta–Release Images". HubbleSite–NewsCenter (19 April 1995).

(75) Ellwood R. S. (1996). "Theosophy". In Stein, Gordon (ed.). The   Encyclopedia   of   the   Paranormal. Prometheus   Books   (pp. 759–766).;   Regal   B. (2009). Pseudoscience: A   Critical   Encyclopedia.

Greenwood (p. 29).; Drury N. (2011). Heaven: The Rise of Modern Western Magic. New York: Oxford University Press (p. 308).

(76) McClelland N. C. (2010), Encyclopedia of Reincarnation and Karma, McFarland.; Juergensmeyer M., Roof W. C. (2011). Encyclopedia of Global Religion. SAGE Publications.

(77) Rita M. G. (1993). Buddhism After Patriarchy: A Feminist History, Analysis, and Reconstruction of Buddhism. State University of New York Press (p. 148).; Flood, G. D. (1996), An Introduction to Hinduism, Cambridge University Press.

(78) Klaus K. (1985). Mokṣa and Critical Theory, Philosophy East and West, 35, 1, 61–71.; Norman E. T. (April 1988), Liberation for Life: A Hindu Liberation Philosophy, Missiology,16, 2, 49–160.; Gerhard O. (1994), La Délivrance dès cette vie: Jivanmukti, Collège de France, Publications de l'Institut de Civilisation Indienne. Série in-8°, Fasc. 61, Édition-Diffusion de Boccard (Paris) (pp 1-9).

(79) Kusum P. M. (1996). Yama, the Glorious Lord of the Other World. Penguin (pp. 213–215).

(80) Taliaferro C., Draper P., Quinn P. L. (2010). A Companion to Philosophy of Religion. John Wiley and Sons (page 640).

(81) Strassman R. (2000). DMT: The Spirit Molecule. Inner Traditions.

(82) Silberberg, M. (2006). Chemistry: The Molecular Nature of Matter and Change, 4th Ed. New York (NY): McGraw-Hill Education (p. 935).

(83) Was King Tut Buried in a Hurry?". History.com.